ERECTILE DYSFUNCTION
MANAGING & OVERCOMING

CONTENTS

INTRODUCTION

Whether you call it erectile dysfunction, ED, impotence, or any number of slang terms, erection problems are something many men have to face during the course of their lifetimes.

But what you might not realize is that erection problems aren't just a matter of getting older. You can prevent and treat most erection problems when you know what they are, how they are caused, and what treatments are available.

Most men avoid getting treatment for ED for these very reasons. Whether it's embarrassment or simply feeling like there's nothing to be done, erectile dysfunction is often ignored when steps can be taken.

Choosing to manage and to treat your ED is a powerful and a courageous step, to be sure. But you can take this step knowing not only that help is available, but it's also going to help you manage your problems and get back the sex life you desire.

In this book, we are going to talk about:

- What ED is and what it isn't
- Possible risk factors
- Potential causes
- How to get tested
- Treatments and therapies that can help
- How to prevent ED in the first place

You might not be feeling too great about yourself right now, but that doesn't need to be the way of your life anymore.

It all starts with knowledge. Here's what you need to know.

WHAT ED IS AND WHAT IT ISN'T

It's interesting that many men just aren't certain what erectile dysfunction is, what it means, and what it doesn't mean.

The easiest definition is that erectile dysfunction is the inability of the man to maintain a strong erection long enough to have sex. That's the simple version.

However, for many men, it can be a bit more complicated.

FACT: ED IS NORMAL

First of all, you need to understand that having troubles with impotence occasionally is completely normal. Every man will face this problem at some point in their lives.

Erectile dysfunction is often defined as:

- An inability to maintain an erection
- The tendency to maintain only brief erections
- The inability to be consistent in erections

It's when it becomes more of the norm than a fluke that you want to become concerned. If you've noticed that your erections aren't as hard as they used to be or that you're having troubles becoming firm, this isn't just something to accept – this is a problem to address.

Though ED can be seen as normal, it can also be avoided with the proper information.

FACT: ED IS NOT TABOO ANYMORE

There used to be a time when impotence or erectile dysfunction was a taboo topic – but this is not the case anymore.

With more ED drugs available and regular commercials and advertisements for ED medications, erectile dysfunction has become another condition that modern medicine has been entrusted to treat.

You shouldn't feel ashamed to talk to your partner or to your doctor about erectile dysfunction. Creating open communication about ED and solving the problems of ED will allow your personal relationships a chance to blossom. And you will feel stronger and more self-assured as you take control.

In the end, you're helping yourself and your partner get more out of life as a result.

FACT: ED IS NOT PREMATURE EJACULATION

Do not become confused by the other sexual problem that are out there – like premature ejaculation. Many men see this as impotence too, but true erectile dysfunction focuses solely on the erection and whether it can be maintained or not.

Premature ejaculation occurs when a man can still have an erection, but he simply has an orgasm too quickly for his and his partner's liking.

While this is an issue, it is not erectile dysfunction.

That said, there are some people who face premature ejaculation who then have troubles with ED. But it's generally not seen the other way around.

FACT: ED IS NOT AS COMMON AS YOU THINK

One thing to keep in mind is that erectile dysfunction is not all that common at all. While there are many men who suffer from this condition, here are some statistics to keep in mind:

- By 1999, about 22.3 visits out of 100 to one's primary care doctor by men were for erectile dysfunction and other sexual problems – NAMCS.

- 5% of all men around the age of 40 will experience ED.

- 15 – 25% of men around the age of 65 will experience ED troubles.

But when you stop to look at these figures, you can see just how low they actually are.

The physical process that an erection goes through is one that many men do not understand, which often leads to misunderstandings about ED and how it happens.

FACT: ED HAPPENS BECAUSE OF BLOOD FLOW ISSUES

Here's what you need to know.

Within the penis are two chambers filled with a spongy tissue. This tissue includes things like veins, arteries, smooth muscles, and other tissues.

When a man is stimulated visually or emotionally, the brain tells the muscles of these chambers to relax, thus letting the penis fill with blood. This blood flow will cause the penis to become erect.

During the orgasm, the muscles are contracted, allowing the blood flow to reverse and release the actual erection.

The problem with ED is that something happens to disrupt the blood flow to the penis, causing the erection to be less than firm or to be nonexistent. This can be due to a physical issue or an emotional issue – we'll cover that in a moment.

FACT: ED IS TREATABLE

What you might not realize (yet) is how many treatments and therapies are available right now. While we're all pretty familiar with the little blue pill, this is just one of the many different ways you can help your body achieve an erection.

Even if the first treatment doesn't work for you, there are other ways to manage ED and to ensure you have a satisfying sex life.

You need to remember that there are treatments like:

- Medications
- Relaxation techniques
- Supplements
- Surgeries
- Lifestyle changes

All of these or just a few of these can work together to ensure your penis is ready for that 'right moment.'

<u>FACT: ED IS NOT YOUR FAULT OR ANYONE ELSE'S FAULT</u>

Something you need to keep in mind right now is that erectile dysfunction isn't your fault in any way, shape, or form.

In the end, ED is something that is treatable or manageable, but it is not a condition that responds to blame. While there many be lifestyle factors that play into your risk of ED, chances are good that you didn't know it before.

And now you will.

You can stop blaming yourself and others for your erectile dysfunction and just focus on changing your life.

This is an issue that simply just happens from time to time. You aren't necessarily able to control it or to stop it when you're unaware of how you might help the situation.

Something else you might want to keep in mind as you are looking for help with erectile dysfunction is to realize that this issue isn't something that's hopeless. Even though you've probably begun to understand this, it's something the bears repeating.

Right now, realize that you can change your sexual problems into sexual successes. Now, let's get started.

COULD I BE NEXT?

Whether you've already had troubles with erectile dysfunction or not, you might want to think about whether you are putting yourself at right now. Chances are good that you are healthy and not at risk, but if any of these factors sound like you, it might be time to change the way you live.

And this will help you keep your sex life intact.

GETTING OLDER

While it's true that not all men are going to have troubles with erectile dysfunction as they get older, there does seem to be a link between age and troubles with erection function.

This might be partially due to other causes – lack of physical fitness, other illnesses and diseases – that are also linked with old age, but there are also some concerns that age related hormone drops might also cause the troubles with erectile dysfunction.

According to the Mayo Clinic, up to 80% of all men 75 and older will have troubles with erectile dysfunction. But this doesn't mean that men of that age need to simply give up on their sex lives.

There many just be some simple changes that need to be made to ensure that prime sexual function is maintained.

MEDICATIONS

Whenever you are taking a medication, there are side effects – for any medications. And since all of our body chemistries are different, it can be impossible to decide whether someone will react to particular medications in a certain way.

What happens with these medications is that they can inhibit the flow of blood to the penis, which can then cause troubles with stimulation and maintaining that erection.

However, changing medications can often help the patient see improvement in their erections and in their overall sexual performance. Of course, if you are on any medications right now, this doesn't mean you should just stop taking your prescribed medications.

ILLNESSES AND DISEASES

If you have a chronic illness or disease, this too can be affecting the way your body sends blood to the penis. Any troubles with these systems can lead to potential erectile dysfunction troubles:

- Arteries
- Veins
- Nerves
- Kidneys
- Liver

- Lungs
- Heart

As you can see, nearly every part of your major organ systems can in some way affect the way you maintain your erection. In fact, having ED can sometimes be a signal that something bigger is wrong with your body which is why men should get themselves checked out by their doctors.

On the other hand, a lack of testosterone can also put you at risk for erectile dysfunction as can plaque build-up in your arteries as that can hinder blood flow throughout your body.

EMOTIONAL TROUBLES

Since the brain is in control of the rest of your body, when you have troubles with stress and with depression, you might be putting yourself at risk for troubles with erectile dysfunction. Though it seems as though your brain isn't interconnected with the blood flow to your penis, this is not entirely accurate.

When your brain is affected, it can cause troubles with the rest of your body. After all, if your brain isn't able to focus on the needs of the body, some functions are going to suffer.

If you're prone to troubles with your emotions, you might have a greater risk of ED.

DRUGS AND ALCOHOL

If you have a substance abuse problem, that can put you at a higher risk of erectile dysfunction. By hindering the body and the mind, you are enhancing the troubles in your arteries and veins, which can cut off the blood flow to your penis.

This is especially the case when you've been using drugs and alcohol for a long period of time. You may have done irrevocable physical damage to your body, which may result in permanent problems with ED.

This includes illegal drugs as well as addictions to legal pharmaceuticals that have not necessarily been prescribed by your doctor.

HIGH BODY FAT

If you've overweight, you might also be putting yourself at risk for troubles with erectile dysfunction. Since excess pounds can strain your system and can cause your body to work harder than it should, you might have troubles with getting the blood flow to your penis.

Since the excess body weight also can cause health problems, this can also become a risk factor for erectile dysfunction.

Linked to the risk of being overweight is the idea of a metabolic disorder. If you are suffering from this type of issue, it can lead to troubles with your blood pressure, cholesterol as well as with your ability to process sugar. Men who have larger waistlines are at risk for this type of issues.

INJURY

If you've ever had a penis injury, you can also be at risk for erectile dysfunction, obviously. But this risk also extends to any lower body injuries that might be connected to the function of the penis as well. If you're unaware of any injuries, you might want to talk to others who may be able to give you an idea of what past events may be likely to contribute to ED troubles.

Even minor injuries can cause major problems, so when talking to your doctor make sure to discuss any intentional or unintentional penis injuries.

SURGERIES

If you've ever had surgery that involved your genital area or that involved the arteries and major organs of your body, this might put you at risk for erectile dysfunction as well. Whenever you disrupt or change the way your body works, you put the rest of your body at risk for future problems as well.

SMOKING

Men who smoke also seem to be at a greater risk of developing erectile dysfunction troubles. Since smoking constricts the blood vessels, there can be troubles with blood flow in the penis, which can then lead to greater troubles with erection maintenance.

Smoking also leads to other physical problems which might also increase your chances of ED.

CERTAIN EXERCISES

Anything that places a lot of pressure on your penis and your scrotum can be problematic for the maintenance of erections. For example, men who bike for long distances can be at a higher risk of erectile dysfunction issues.

This is not to say that exercise is bad, just that some exercises may need to be done in moderation or with protective gear to help the biker protect their sexual function.

Knowing these risk factors can help you make healthier choices in your life which will allow you maintain your ability to have erections for a long period of time.

At the same time, you need to keep in mind that while some men might have all of these risk factors, they may also not have any troubles with their erections.

So while this list seems to factor in the consensus of the medical world, it might not apply to your situation. Some men just have erection problems without any risk factors.

HOW DID THIS HAPPEN?

What most men don't realize is that there are a number of different causes that are associated with erectile dysfunction – no one diagnosis fits all.

In order to select the best treatment possible for you and for your erectile dysfunction issues, you need to make sure (along with your doctor's help) that you are finding the true cause.

PHYSICAL CAUSES

Here are some of the physical causes associated with erectile dysfunction:

- **Alcohol and tobacco use**
  Whenever you use alcohol and tobacco, you begin to damage the body. The more you drink and the more you smoke or dip, the more damage you do. As you continue to alter your body in this way, you can begin to create other health issues as your body simply can not repair the damage you've done. Now, the good news is that once you do stop the damage-causing behavior, you can generally heal your body and get back on track. But you will need to stop the unhealthy behaviors.

- **Atherosclerosis**
  Hardening of the arteries can happen for a number of reasons. You might be getting older, and this is a natural result of aging. If you aren't taking care of your body, you might also have troubles with

artery hardening. Fatty foods, too much sugar and a lack of exercise can all contribute to this condition as well. Like alcohol and tobacco use, much of this damage can be undone as well as controlled by other medications.

- **Brain or spinal-cord injuries**
When your brain is unable to send the message to your body to let blood into the chambers, you will have troubles with erections. Thus, if there has been damage to the brain or to the spinal cord, this can cause erectile dysfunction. In some cases, however, this damage can be reversed and proper function can be restored.

- **Diabetes**
If you have been diagnosed with diabetes, you already know how much your life changes once you have that illness in your life. But what you might not realize is that not only can unregulated blood sugar levels cause your health to deteriorate, but they can also cause erectile dysfunction. By causing troubles with your arteries and with your body's ability to process sugar, you can notice a distinct change in the way your penis performs. You can create irreversible nerve damage when you do not use insulin in the proper manner and when you do not control your diet. However, managing your diabetes can help to correct these troubles since regulated blood sugar levels will not cause damage to your body.

- **Exhaustion**
Your body needs several things to take care of itself – a good diet, plenty of exercise, and rest. When you don't allow your body to rest, you can cause damage to your body systems as well as prevent normal function of your organs. When you don't allow your body to rest at all, you will see your erections become fewer and fewer, while also noticing that you have additional health problems as a result.

Your body needs about 7 hours of sleep each night to rest and to recuperate. Without this amount of sleep (or more) on a daily basis, ED can ensue.

- **Heart disease**

  Your heart allows your body to move blood and essential things like nutrients and oxygen to all parts of your body. When your heart isn't beating correctly or the blood isn't flowing through arteries, you will have troubles with your erections. There are many different types of heart disease that can cause troubles, but what's even more upsetting at times is that the medication used to treat heart disease can also lead to ED problems. Thankfully, there have been advances to keep you healthy and happy without sacrificing your sex life in the process.

- **Hypertension**

  Also known as high blood pressure, hypertension is a condition that affects many more people than they realize. Since there are generally no symptoms, men can go for years without knowing they need to control their blood pressure. Each time the heart heats, it pushes blood to the various parts of the body. When your heart has to push too hard, it causes pressure on the walls of the arteries. This can lead to a thinning of the arteries as well as hardening. All of these outcomes can lead to troubles with your erections.

- **Hypogonadism**

  Some men might notice that their testicles either suddenly get smaller or they slowly seem smaller than they used to before – this is called hypogonadism. When this happens, it's actually a sign that the body isn't producing as much testosterone as it should, which can led to troubles with erectile dysfunction. If you notice this kind of issue, it can also be indicative of other medical troubles, so this should not be ignored.

- **Liver failure**

  The liver is an organ that is used to process toxins and other substances in the body. Also the largest gland, the liver produces bile that is then used to digest foods so they can be used for nutrition. Without your liver, you would only live about 24 hours since it is necessary to break down red blood cells, remove toxins, and to manage various levels of nutrients in the body. When your liver isn't working, it can cause damage to other parts of your body, thus leading to troubles with erections.

- **Kidney failure**

  Your kidneys used to flush toxins from your body as well. When your kidneys aren't working, the toxins can go back into your body and cause multiple organs to fail. This in turn can lead to troubles with erectile dysfunction as the blood flow to the penis may be compromised.

- **Multiple sclerosis**

  Those who suffer from MS might have troubles with nerve function in their bodies, so it's no wonder that this can then affect the erection process. When your body is unable to send the proper signals, this halts many processes, including erections. In addition, MS is a disease in which the body's own immune system attacks itself, causing the nerves to become damaged permanently. It can also lead to troubles with pain, which can also cause erectile dysfunction.

- **Parkinson's disease**

  People who suffer from Parkinson's disease also have troubles with nerve function as the disease attacks the nerves and slowly causes them to degenerate. This leads to troubles with movements and with feeling, which can also lead to troubles in the genital area, causing erectile dysfunction.

- **Peyronie's disease**
  This disease occurs when the penis is unable to maintain a straight curvature outward. The penis' connective tissue is damaged and lesions can form, causing sex to become painful and sometimes awkward. For some men, a normal sex life is possible, but since many men experience significant pain, it can lead to troubles with erectile dysfunction.

- **Stroke**
  As a stroke causes a blood clot in the brain, there might be damage from this process that causes blood flow to be interrupted to other parts of the body as a result. In addition, strokes that cause permanent brain damage will also cause the penis to have troubles achieving and maintaining an erection. Often, correcting the blood flow will help, but brain and nerve damage can be permanent in some rare cases.

- **Some types of bladder or prostate surgery**
  Suffice to say that if you've had surgery in or around the penis area, this can cause significant troubles with your ability to have and to maintain an erection. This might be something you knew about before the surgery, but it can also be slightly surprising to some men as they didn't think anything went wrong with the process. And nothing may have been wrong with the surgery, but changing around anything in the body can have other effects.

- **Treatments for prostate cancer**
  When you've had prostate cancer, the treatments that sent your cancer away may also have sent your erections. From radiation to

surgery, the blood flow to your penis may have been compromised, leading to erection problems.

Physical causes are some of the most common causes of erectile dysfunction in men of any age. When there is physical damage to the penis, it makes sense that you might suffer from erectile dysfunction.

But the good news is that physical causes tend to be easier to handle than other causes of erectile dysfunction, like those which are emotional in nature.

EMOTIONAL CAUSES

Emotional and psychological causes of erectile dysfunction are a bit trickier to diagnose and thus treat because many don't make the connection between the two.

Although it seems reasonable that being depressed or anxious might cause you to have troubles with your erections, many men think that ED can only be caused by physical symptoms – but this is not the case.

And here are some emotional causes that can lead to erectile dysfunction:

- **Stress**
  Stress is one of the top reasons why seemingly healthy men have troubles with erectile dysfunction. When your mind is focused on too many other things, you can have troubles focusing, which can and will upset functions in your body. Making your body work too hard with stress can lead to high blood pressure, poor food choices, issues with

weight gain, etc. All of this can add up to erectile dysfunction. However, when you remove the stress, you can remove the problems.

- **Fatigue**

  Being tired in your life can also lead to troubles with erectile dysfunction. If you're constantly running yourself ragged with too many things on your To Do list and you're simply not doing what it takes to rest yourself emotionally, you will have troubles focusing long enough to maintain an erection. Emotional fatigue can present itself as troubles concentrating, a lack of motivation, and simple problems with keeping up.

- **Anxiety issues**

  Those who already have troubles with anxiety like those who suffer from panic attacks and other panic disorders can find that when their anxiety goes uncontrolled, they can suffer from erectile dysfunction. The scattered thoughts and the physical symptoms like a pounding heart, troubles breathing, and sweaty palms may not seem like major concerns, but they can add up to making a man unable to function sexually. In addition, as these symptoms increase and the erectile dysfunction occurs, a cycle of fear can also begin, leading to more anxiety, and more ED.

- **Depression**

  There is a stereotype that men don't or shouldn't be depressed in their life, but this is not only incorrect, but also ridiculous. Many men have troubles with depression, whether caused by physical chemical imbalances or by events that take place in their lives. When you're sad about your life, your body responds in kind, keeping you from performing sexually as you might like.

- **Communication issues**

  If you are having troubles communicating at work or in your relationships, this can also cause troubles with ED. The frustration that can build up in these situations will cause you to not only tense your entire body, but it will also make your blood pressure rise. Without learning how to handle your communication issues, you can continue to have troubles with ED.

- **Troubles at work**

  Since many men equate their work and their career with their self esteem, troubles that arise in the workplace can cause erectile dysfunction troubles as well. When you have troubles at work, you will have more anxiety, more stress, more fatigue, etc. All of this adds up to troubles in your entire life.

- **Relationship conflicts**

  Within your relationship with your sexual partner, you need to have balance and peace in order to have a satisfying sex life. If you're constantly fighting, this is nearly certain to cause occasional troubles with ED. And if you and your partner are constantly at each other's throats, you can continue to suffer from erectile dysfunction until the relationship is fixed.

- **Self esteem issues**

  Men who are not happy with the way they look or the size of their penis may also have troubles with erectile dysfunction. Feeling as though you're not going to be 'good enough' or that you're not an attractive man not only causes your body to tense up, but it can also create negative feelings that go along with your sex life. Together, those feelings can create long term ED if not addressed.

No matter what the cause of your erectile dysfunction, there are ways to handle it and to reverse your troubles in most cases.

Knowing what the potential causes may be can help you and your doctor work together to find the right treatment for your particular case. Take some time to look through these causes to see what sounds like you and your life.

TESTS TO HELP DIAGNOSE AND TO CHOOSE THE BEST TREATMENT

When you recognize you might have a problem with erectile dysfunction, you need to begin to consider what steps to take next.

Most men and their partners might decide to try to 'fix' things on their own through trying new sexual positions and trying new things to somehow stimulate their sex life.

This is a good way to begin. Often, men can reenergize their sex lives on their own, without any medical intervention.

But just as trying to conceive a child naturally isn't possible for some couples, sometimes erectile dysfunction just isn't something you can take care of on your own.

SIGNS YOU NEED MEDICAL ATTENTION

Here are some signs you need to find medical attention:

- **Avoidance of sex because of ED**
  When you're beginning to avoid sex because of your troubles with

erectile dysfunction, it's time to start talking to your doctor. You should never feel as though your sex life has come to a halt because of this often easily treatable and manageable problem.

- **Repeated attempts at erections without success**
  If you've tried on your own and with a partner to regain your ability to control your erections, but have not had any luck, it's time to find medical attention. When things that used to work in the past just aren't working anymore, it can be a sign of true erectile dysfunction and the need for intervention.

- **Complete inability to maintain an erection**
  If your erections are simply nonexistent, it's time to start looking for outside help. Whether this happens every single time you have sex or every single time you try to masturbate – or both – you need to check to make sure there isn't something physically wrong with you and your health.

- **Pain associated with sex**
  When sex becomes painful when you are trying to maintain and to control an erection, leading to ED or as a result of ED, it's time to find help.

- **Relationship troubles as a result of ED**
  If erectile dysfunction is causing your relationship to begin to suffer, it's time to talk to someone about helping your erectile dysfunction.

Many men have troubles with talking to their doctors about this issue because they feel as though it shouldn't be an issue. Ever. But since erectile dysfunction is a part of many men's lives, it's time to dispel some of the societal ideas that surround ED and other sexual dysfunction.

MYTHS ABOUT ED

MYTH #1: MEN SHOULD NEVER HAVE THESE SORTS OF PROBLEMS

You've already learned that erectile dysfunction is fairly common and certainly something that many men will face at one point or another in their lives. However, having erectile dysfunction problems does not make a man any less of a man.

MYTH #2: MEN SHOULD JUST KNOW HOW TO HAVE SEX

Men are often given the role of being in charge of their relationship's sex life, so they are often looked at as having all the answers when it comes to sex and controlling the sexual relationship.

As a result, it's thought that all men simply know how to have sex and how to correct issues when they come up. Thus, when something like erectile dysfunction comes into their lives, it's their fault for not being 'man' enough to know how to have sex. False. False. False.

Men, just like women, learn to have sex as they have romantic partners and through their own self-exploration. One is not born with the innate ability to have the perfect sexual relationship.

MYTH #3: MEN ALWAYS WANT AND CRAVE SEX

Though the media and many other sources might have you believe otherwise, men are not necessarily always looking for sex. This is something that can cause a man to feel pressure just as much as a woman might.

And this ideal can often lead to shame about problems with erectile dysfunction. After all, if you're having problems, you might not feel like having sex, which then complicates the 'ideal' that all men want and need to have sex.

MYTH #4: IT'S EMBARRASSING TO TALK TO SOMEONE ABOUT ERECTILE DYSFUNCTION

At first, sure, erectile dysfunction might be a little embarrassing to talk about, but that doesn't mean it should be avoided.

There are plenty of ways to approach the subject in a simple and clear way without feeling as though you're being judged. Every man understands the fears of this dysfunction – you're not alone.

MYTH #5: ERECTILE DYSFUNCTION IS JUST THE MAN'S PROBLEM

As you begin to explore the process of healing erectile dysfunction, you might begin to think it's just your fault or that it's just something you need to deal with on your own. This is also false.

Erectile dysfunction affects both members of a relationship, so both persons should be involved in the healing process.

FINDING A DOCTOR FOR YOUR ED QUESTIONS

The first thing you need to do when you feel you have troubles with erectile dysfunction is to find a doctor that will be able to help. For most men, this will begin with talking to your general care doctor to see what they might have to offer in terms of advice and guidance.

You want to make sure, of course, that you feel comfortable enough with your doctor to be able to discuss your symptoms and your concerns openly. If you feel your doctor is unable (or unwilling) to listen to you, you might want to ask other male friends who they might recommend for further care.

The process of erectile dysfunction healing begins with this primary care doctor and here's what you can expect to be asked.

WHAT IS YOUR FAMILY HISTORY?

While family history does not necessarily tell you whether or not you will have troubles with erectile dysfunction, it can give your doctor an idea of health issues that you might be at risk for.

If you have a family history of heart or kidney troubles, for example, your doctor might want to do tests on these systems of your body to make sure your erectile dysfunction isn't an early symptom of serious problems.

Make sure to be honest about your family history, especially if you're not sure about certain members of your family. Before you head in for your appointment, you might want to check with your parents to see what kinds of medical issues they have had and what they were especially dealing with at your current age.

HOW'S YOUR HEALTH?

Here's the part where it is crucial you be upfront and honest with your doctor. You need to talk about all of your health habits, being honest and forthright.

When you are asked if you smoke, you need to be truthful.
When you are asked about drinking, you need to be truthful.

You might want to make a list of all of your recent health concerns to make sure you discuss them with your doctor right from the start. This will help you create a timeline of the events that led up to your erectile dysfunction to see if there are any signs or related issues that might give the doctor answers.

Remember, nothing is too small to mention.

WHAT ELSE IS GOING ON IN YOUR LIFE?

A good doctor will also be sure to note whether or not you've been going through major stressful situations. While the doctor may not be able to help you with these issues directly, knowing they might be factors in your erectile dysfunction can help to direct you to specific caregivers, like therapists.

Again, be honest as you talk to your doctor about your emotional state of mind. Your doctor needs as clear a picture as possible about what may or may not be contributing to erectile dysfunction.

WHAT HAVE YOU TRIED TO ALLEVIATE YOUR ERECTILE DYSFUNCTION?

If you have been dealing with erectile dysfunction for a while now or you are suffering a recurrence of erectile dysfunction, you will want to talk to your doctor about what you have already tried on your own to get things working again.

Try to be as specific as possible to show the doctor what they can rule out in terms of potential treatment options as well as to help the doctor see what might be signaling the actual cause of the erectile dysfunction.

HAVE ANY METHODS OF ERECTILE DYSFUNCTION TREATMENT WORKED IN THE PAST?

If you have been successful with any of your erectile dysfunction methods of treatment in the past, be sure to share this as well. For example, if you notice you don't have troubles with erectile dysfunction when you masturbate, but you do with your partner, that's a significant cause for concern.

TESTS YOU WILL NEED TO TAKE

When your doctor is trying to determine the root cause of your erectile dysfunction, you may need to undergo a battery of tests in order to get answers.

While some doctors will simply prescribe treatments until they find one that works, having these tests will help to narrow down the field of possible treatments before you start any regimens, thus speeding up the healing process.

Here are some of the tests you might need to take:

GENERAL CHECK UP

This might include everything from a blood pressure check to a weight check. These simple tests are painless and they will help your doctor get an overall impression of your health before they go further.

BLOOD TESTS

Checking your blood levels can give your doctor a clear understanding as to whether you have troubles with your blood lipid levels and your overall health. These tests will measure everything from your HDL and LDLs to your liver enzymes and creatinine levels.

If part of your blood work is off, it can not only signal a potential cause for your erectile dysfunction, but it can also signal other health problems which may need to be addressed.

URINE TESTS

Checking your urine will help your doctor determine if your kidneys are functioning properly, while also checking to be sure you don't have blood sugar issues.

HORMONE TESTS

If your erectile dysfunction also comes with problems with libido, you might need to have your testosterone levels checked.   This is called a free testosterone test and it can be run on the blood that was taken for other tests.

ULTRASOUNDS

By using an ultrasound wand, your doctor can also measure the amount of blood flow to the penis to see if there are any blockages or if there is any trouble in the artery structures.

The ultrasound is a painless procedure in which the wand is placed onto the penis and genital area and then moved around to measure different rates of blood flow.

At times, this test will be performed in the presence of a certain medication and then in the absence of a medication to measure changes in blood flow – if any.

NEUROLOGICAL TESTS

When your doctor is concerned about nerve damage, you might need to have these sorts of tests. In most cases, all this requires is a physical examination in which your doctor will palpate the penis and the genital area to see if there are any nerves that aren't functioning correctly.

PSYCHOSOCIAL TESTS

When your doctor suspects your erectile dysfunction might be caused by emotional difficulties, you might be given a questionnaire to fill out about your feelings about sex and about your partner.

Your partner might also be given this test to see how you both relate to each other sexually and in response to this erectile dysfunction issue.

DICC – DYNAMIC INFUSION CAVERNOSOMETRY AND CAVERNOSOGRAPHY

After injecting a dye into the penis, the doctor can then watch the blood flow through the penis to see whether there are obstructions or not. A urologist is a specialist that might perform this tests and it only requires local anesthetic to prevent pain and discomfort.

NOCTURNAL TUMESCENCE TEST

If other tests are inconclusive or your doctor thinks your erectile dysfunction might be caused by non-physical issues, you might take the nocturnal tests.

This is actually a test you can do on your own as well.

By wrapping a piece of perforated (non-sticky) tape around the penis before you go to bed, you then go to sleep as normal and check to see if the tape has broken by the morning.

If the tape is broken, this might be a sign that you don't necessarily have troubles with the physical creation of an erection, but you might have other non-physical issues to address.

While this can seem like a long and complicated process of testing and checkups, all of these steps will help you and your doctor narrow down the possible causes of your erectile dysfunction so that you get the best possible treatment.

Beyond your family doctor, you might also be referred to a urologist who specializes in dealing with the physical issues surrounding the penis and other related areas and organ systems.

TRADITIONAL AND
ALTERNATIVE TREATMENTS

Once you and your doctor have determined the cause of your erectile dysfunction, it's time to start treating the problem.

Today, there are many more treatments than ever available for erectile dysfunction before. This also provides you with many more decisions to make as you navigate the erectile dysfunction healing process.

Becoming informed about your options it the key to making sure you have all the information you need while also looking to see what you would be willing to do as a part of your treatment.

Here are the major categories of erectile dysfunction treatments, what they include and how these treatments are administered.

MEDICATIONS

The most popular and the most prescribed treatment for erectile dysfunction is medication. This generally allows the man to take control of his erections immediately, while also providing privacy in terms of solving the problem.

The three main medications for treating erectile dysfunction are:

- Tadalafil – Cialis
- Vardenafil – Levitra
- Sildenafil – Viagra

Each of these medications works to improve blood flow to the penis and the medications are administered orally in various dosages

HOW THESE MEDICATIONS WORK

These medications work by enhancing the effects of nitric oxide in the penis. This naturally occurring substance helps to relax the muscles of the penis in order to restore blood flow.

Once the medication is taken, an erection is not always immediate, however. These medications will provide the right circumstances, but the man will still need physical or mental stimulation to encourage the erection to start.

The good news it that these medications often help you no matter what the cause of the erectile dysfunction. So, for most men, these are the first step to erectile dysfunction freedom.

The bad news is that there are side effects to these medications, just as there are for all medications.

POSSIBLE SIDE EFFECTS

Some men experience headaches from these medications as they do dilate the blood vessels. Those who have heart problems can also cause more troubles with their blood pressure, so your doctor will need to keep this in mind when choosing the right formulation for you.

This is why it's imperative that you talk with your doctor before taking any of these medications. Self-medicating with someone else's prescription can be fatal.

Men who take these medications will not be able to use these erectile dysfunction medications:

- Blood thinners – ex. Coumadin, warafarin, etc.
- Nitrate drugs – nitroglycerin, etc.
- Alpha blockers

Patients with the following conditions will also want to talk to their doctors about whether these medications are right for them:

- Past history of strokes
- Uncontrolled diabetes
- Low or high blood pressure

Some patients may also experience long and painful erections that have troubles subsiding without medical intervention. If a man experiences an erection lasting more than 4 hours, they should see their doctors.

With the three different formulations, some of these medications might work for some men, while others work better for other men. There is no 'best' medication for all cases, so you may end up trying each one or just one medication at different dosages.

Make sure to take these medications as directed and talk to your doctor about the effects you experience.

Another medication that can be helpful to men is Prostaglandin E (alprostadil). This erectile dysfunction medication is hormone-based and is used to relax the penis muscle tissue in order to help blood flow for an erection.

This medication can be administered in two different ways:

- **Urethral administration**
  By taking a disposable applicator and inserting a small suppository of alprostadil into the tip of the penis, this medication will be absorbed into the muscle tissue. Since the medication does need to be inserted several inches into the urethra, this can be painful for some men, which can make this an undesirable method of treatment. Other side effects include bleeding, fibrous tissue formation, and dizziness.

- **Needle injection**
  Instead of inserting the alprostadil into the penis itself, a fine needle injection can eliminate the pain associated with getting this hormone into the organ. This method will produce an erection within 5 to 20 minutes, and the erection can lest about an hour. This tends to be a highly effective treatment, thought it can be expensive and for men who don't like needles, a little frightening.

Some men who test positively for testosterone deficiencies will have to undergo hormone replacement therapy of sorts, as determined by their doctors and the level of the deficiency. This can include medications as well as injections until the levels are returned to normal.

PENIS PUMPS

To help encourage blood flow to the penis, penis pumps are often used in erectile dysfunction treatment as well. This battery powered pump is placed over the penis and then the extra air around the penis is sucked away.

As a result, blood flow is enhanced to the penis area, created an erection.

But once the erection is created, you will need to place a tension ring around the base of your erection to ensure it can be maintained for the duration of your sexual experience.

Once you have had sex, you can then remove the tension ring.

For those men who do not want to take medications and who are unable to take medications, this can be a simple and often very effective treatment for erectile dysfunction.

While it can seem a little awkward and even comical at first, this is a tested and proven method of erection assistance. You can find these penis pumps online and in high quality sex shops to buy them in a discreet manner.

These penis pumps can be expensive, but in relation to the cost of medications and other treatments, they're quite reasonable. However, you might need to buy several pumps before you find one that works well for you.

If you find you have troubles maintaining an erection more than you have troubles achieving one, you might want to check out cock rings at the same sorts of retailers. Just as you place that tension ring at the base of your penis after using the pump, these rings are placed in the came location after you have achieved an erection on your own, helping you to maintain your hardness for the duration of intercourse.

SURGERY

When you have significant problems with the nerves or the arteries in your penis, you might need to have surgery to fix these issues. But take heart, this is rare and not the first thing a doctor will suggest when you have troubles with erectile dysfunction.

Surgery is often the best course of treatment when you've had some sort of injury to the penis area or when you have suffered from cancer that required a surgery during the treatment.

Another type of surgery that is more common, though still not something to consider before less invasive treatments, is the penile implant.

This is a surgery in which the surgeon places a device into the shaft of the penis in order to create an erection for as long as the man decides he wants to maintain that hardness. The device is inflatable, so the man has complete control over his erections permanently.

The cost of this surgery can be quite high and not all insurance companies will cover this treatment, so it might be best left as a last, last resort.

In addition, as with any surgery, you need to remember that there is a chance of infection, excessive bleeding, and other problems as a result.

THERAPY AND COUNSELING

Whether your doctor feels the cause of your erectile dysfunction is simply physical or entirely emotional, it can be helpful to have some sort of counseling as you manage ED.

In terms of treatment, therapy can help you learn to deal with stressors and conflict in your life that might be cutting off blood flow to your penis.

COGNITIVE THERAPY

When you undergo cognitive therapy, this will work on breaking down any fears and hesitations you might have about sex in your life. This is a program in which you will have specific goals to achieve as well as specific steps in which to achieve them.

These types of therapy arrangements are for limited time periods, which can work well for people who don't have a lot of mental health coverage on their insurance plans and for those who don't want to spend their entire lives in therapy.

THERAPEUTIC MEDICATIONS

Some psychologists may prescribe medications to help with severe depression or anxiety, but these medications can also lead to ED, so you will need to be monitored closely to see if the benefits of healing your mind are in line with healing your erectile dysfunction.

Some patients will be able to take smaller doses of these medications without losing any sexual performance or desire.

SEX COUNSELING

When you begin to realize that your relationship might be the source of some of your erectile dysfunction problems, you might want to head to therapy together. In this way, you can learn how to communicate with each other about your needs, while also learning how to deal with conflict when it does arise.

ALTERNATIVE TREATMENTS

Just as with so many other medical fields, the treatment of erectile dysfunction has also begun to see alternative therapies become standard practice. These treatments may be used in conjunction with other treatments you might have in place, but make sure to check with your doctor before starting any new treatment.

HERBAL SUPPLEMENTS

Before we talk about herbal supplements for erectile dysfunction, there's something you need to keep in mind. There are numerous supplements out that that claim to help with erectile dysfunction, but are actually more harmful than helpful.

When in doubt, stick to the rule of – if it sounds too good to be true, it probably is.

Here are some of the latest erectile dysfunction supplements that are being recommended:

- **Gingko**
  This is a supplement that has been linking with improving blood flow in the brain to improve the memory. Thus, it makes sense that it might also be able to help with blood flow in the rest of the body.

- **Ginseng**
  Considered to be an energy enhancer, ginseng too has been linked with increasing blood flow and helping stamina.

- **L-arginine**
  Just as with the medications that are designed to help enhance the effects of nitric oxide, this is what L-arginine does as well.

- **DHEA**
  A building block of testosterone, this supplement is thought to help those with lagging libido issues.

- **Maca**

  Thought to be another sexual response stimulator, maca has a long history of being used to enhance performance.

Because many of these medications have the same qualities as prescription medications, you will want to talk to your doctor before adding them to your daily routine.

ACUPUNCTURE

The Chinese believe that our body is made up of energy and routes of energy. When one of these routes (called meridians) is disrupted, it can have physical consequences, like erectile dysfunction.

By inserting needles into certain points of the body, an acupuncturist can help to get the energy flowing again, while also allowing you to feel completely relaxed and calm.

Needles aren't for everyone, but this treatment is thought to be highly effective as well as providing minimal risk to your health. Acupuncture also does not interact with medications you might be taking.

However, acupuncture can be expensive and it can require multiple sessions order to see the maximum effectiveness.

TANTRA SEXUAL PRACTICES

Sexual energy is thought to be a part of your body, though many people are out of touch with this energy. By following tantra practices, you can learn how to tap back into this energy to prevent and to treat erectile dysfunction.

When utilizing tantra, you might learn how to breathe into your genital area in order to restore the energy as well as the blood flow.

By learning to be in touch with your body, will begin to understand how your body feels before, during and after orgasm, allowing you to take steps that make sense for your body – i.e. certain physical touches.

This practice also helps couples get a better understanding of their sexual relationship, and it creates a renewed sense of trust and understanding.

For those who might be uncomfortable with sex or who have had troubles with erectile dysfunction, this can help to heal the mind as well as the soul and the body.

WHAT TREATMENT IS BEST FOR YOU?

Choosing the right treatment should come down to a few questions:

- What is the success rate?
- How much will it cost?
- What do I need to do?
- Can I maintain the steps I need to take?

When you consider each of these ideas, you will be able to find an erectile dysfunction treatment that not only helps your erection issues, but that also allows you to continue to experience a healthy and satisfying sex life.

PREVENTING ED IN THE FIRST PLACE

No matter what treatment you might choose for your erectile dysfunction, it's always better to prevent a problem instead of to simply deal with the problem.

Here are some of the ways you can prevent erectile dysfunction from becoming a part of your life, as well as being ways to prevent erectile dysfunction from returning once you have treated the problem.

EMOTIONAL CARE

While there can be many physical causes of erectile dysfunction, there are just as many emotional reasons for having troubles with your erections.

In order to prevent these kinds of issues, you might want to spend some time taking care of your emotional health.

- **Try to cut down on stress**
  While this may seem like an impossible task to tackle, cutting down on your stress is going to help keep you relaxed and calm, which reduces your risk of erectile dysfunction. Try to find ways to cut back on your workload and learn to say 'no' to things you don't want in your life.

- **Communicate with your partner**
  As erectile dysfunction and your relationship with a partner are interwoven, you need to make sure you are communicating with each other as much as possible. This will help you keep things out in the open and it will help you resolve issues before they become bigger problems.

- **Find things that make you happy**
  We all need something in our lives to distract us. Find a hobby or an activity that makes you happy and then make sure to make time for it in your life, outside of your relationship.

- **Learn to take things less seriously**
  Chances are good that you might have some sort of issue in the bedroom again at some time, but instead of thinking it's the end of the world, approach it with humor. You might be surprised at how much easier your life becomes when you take things less seriously.

Taking care of your mental state is something that is difficult to do – but essential for your sex life. Share these tips with your partner too!

PHYSICAL CARE

While it's easy to simply focus on your penis when you want to prevent erectile dysfunction, this is only a part of the equation.

You also need to keep the rest of your body healthy so that your erections are strong and long lasting.

EAT RIGHT

What you put into your body is going to affect the way your body functions. Thus, if you eat things which are harmful, your body will be harmed and you can have troubles with erectile dysfunction.

Thankfully, it's become easier and easier to eat well. Here are some simple tips:

- Avoid refined sugars and simple carbohydrates.
- Avoid high fat dairy and proteins, choose low fat and non-fat options
- Eat at least 5 servings of fruits and vegetables a day
- Choose high fiber carbohydrates
- Consider going 'veggie' a few days a week – avoiding meat products
- Drink water and not sugary beverages or beverages that are artificially sweetened

By simply finding ways to incorporate all of the major food groups into each meal, you will help your body replenish itself and work as efficiently as possible.

From time to time, it's okay to eat a little less than sensibly, but when you want to maintain freedom from erectile dysfunction, the more you can eat right, the better.

You might also want to cut out caffeine or at least cut back on caffeinated products as they can constrict the blood vessels and cause troubles with your blood flow.

EXERCISE

Since erectile dysfunction is a problem of blood flow, you need to make sure you are giving your arteries and veins the most exercise you can – by exercising.

Try to get at least 30 minutes of exercise in each day so as to help keep your heart strong and your body lean. And this doesn't mean you need to be sweating a lot or even pushing yourself to the brink of exhaustion.

Just getting up and exercising for a few minutes here and there will not only help you, but also fit into a busy lifestyle.

REGULAR CHECKUPS

Most of us avoid the doctor as much as we can, but when you want to avoid erectile dysfunction, you will want to stop in for a regular checkup once a year or so.

This will help your doctor spot any problems before they affect your sex life, while also helping you to see what you might want to change in your life in order to be as healthy as possible.

HAVE SEX OFTEN

To further encourage good blood flow in your penis, it's essential that you have an active sex life. That's right; you should have sex in order to continue having sex.

Tell your partner to help you with this step or you can supplement your normal sexual activities with manual stimulation.

The point is to make sure you are keeping the blood flow in the genital area, so that it becomes a natural part of your sexual experience. And the more you practice, the better it will become.

TRY KEGEL EXERCISES

Though once thought of as just exercises that could help women with their sexual performance, Kegel exercises are also great for men who want to maintain their sexual stamina and prevent erectile dysfunction.

To find the PC muscles, just go to the bathroom and stop the flow of urine midstream. The muscles you use to stop the flow of urine are your PC muscles.

You can also insert a finger into your anus and tighten the muscles around your finger to find them as well.

Once you get the hang of this, you can perform your Kegel exercises wherever you are – no urine needed, actually. Just tighten and relax as often as you can during the day to beef up these muscles and prevent erectile dysfunction.

FOLLOW YOUR DOCTOR'S ADVICE

Many people who go to their doctor for physical checkups will end up not following the advice that was given to them. In ignoring this advice, you can actually create more problems than you had when you walked into the office.

Take the time to follow the advice and follow any medication regimens that your doctor prescribes.

If you notice any changes, talk with your doctor immediately. In addition, if you notice that something isn't working or you've having troubles with a certain part of your doctor's advice, call the office to see what substitutions can be made.

CONCLUSION

No matter how long you may have been suffering from erectile dysfunction, this does not have to be the case any longer.

By informing yourself, getting tested and taking the time to consider your treatment options, you will not only break free of erectile dysfunction, but you will also become more in control of your ability to enjoy sex.

While you might not be alone in your struggle with erectile dysfunction, this does not mean that it's any less important that you take the first steps today.

You can overcome ED and you can overcome it now.

Make that call to your doctor today.